I Can Count To 10

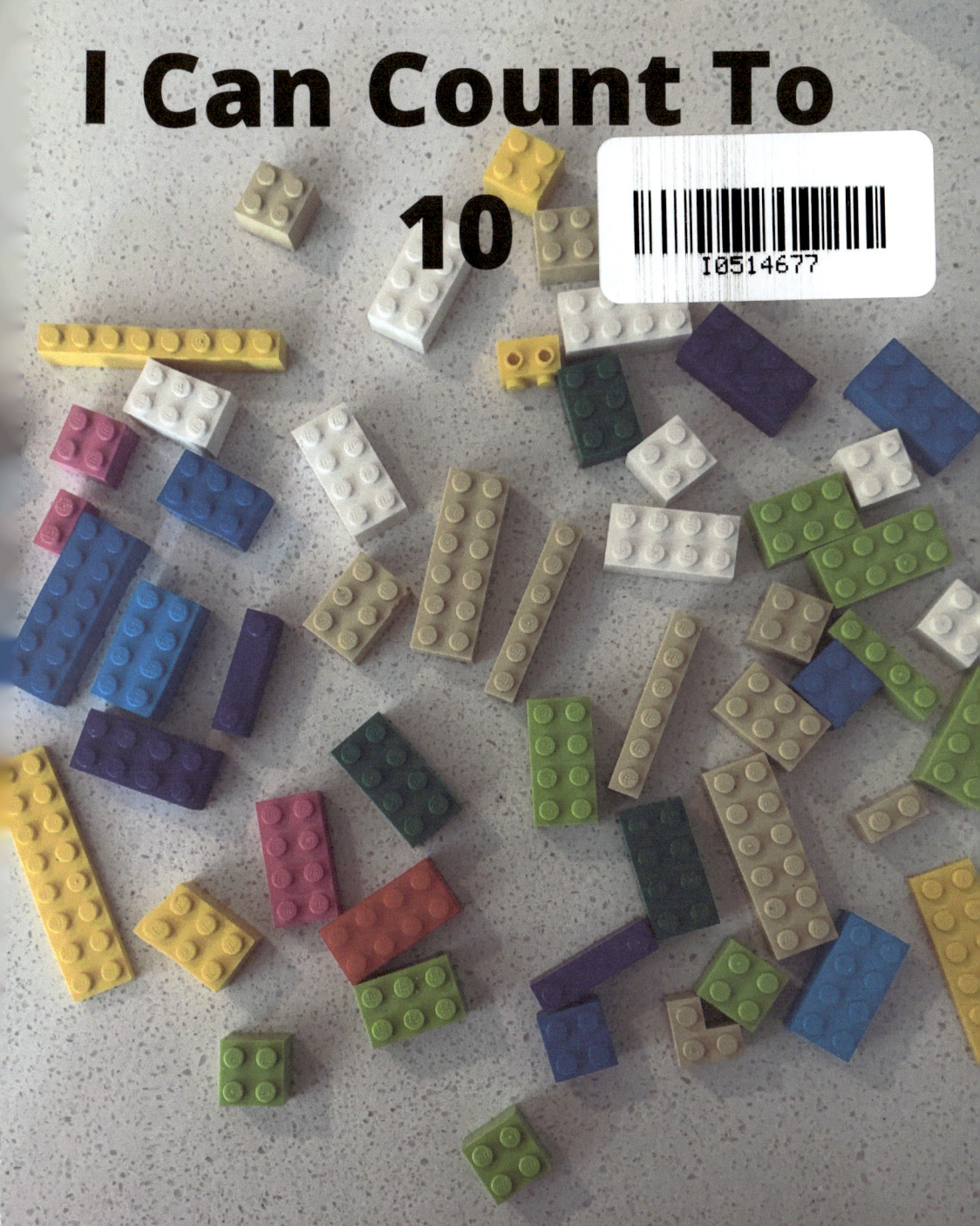

Author and Illustrator
Aimee Ballester

Copyright © 2021 Aimee Ballester

All rights reserved. No portion of this book may be reproduced in any form without permission from the publisher, except as permitted by U.S. copyright law. For permissions contact: Ballesteraimee@gmail.com

ISBN 9798505357323

Dedication

I dedicate this book to kids that are learning how to count. I dedicate this to kids that are struggling and need help to count. I hope by this book and the images of the legos help you out.

ONE

TWO

2

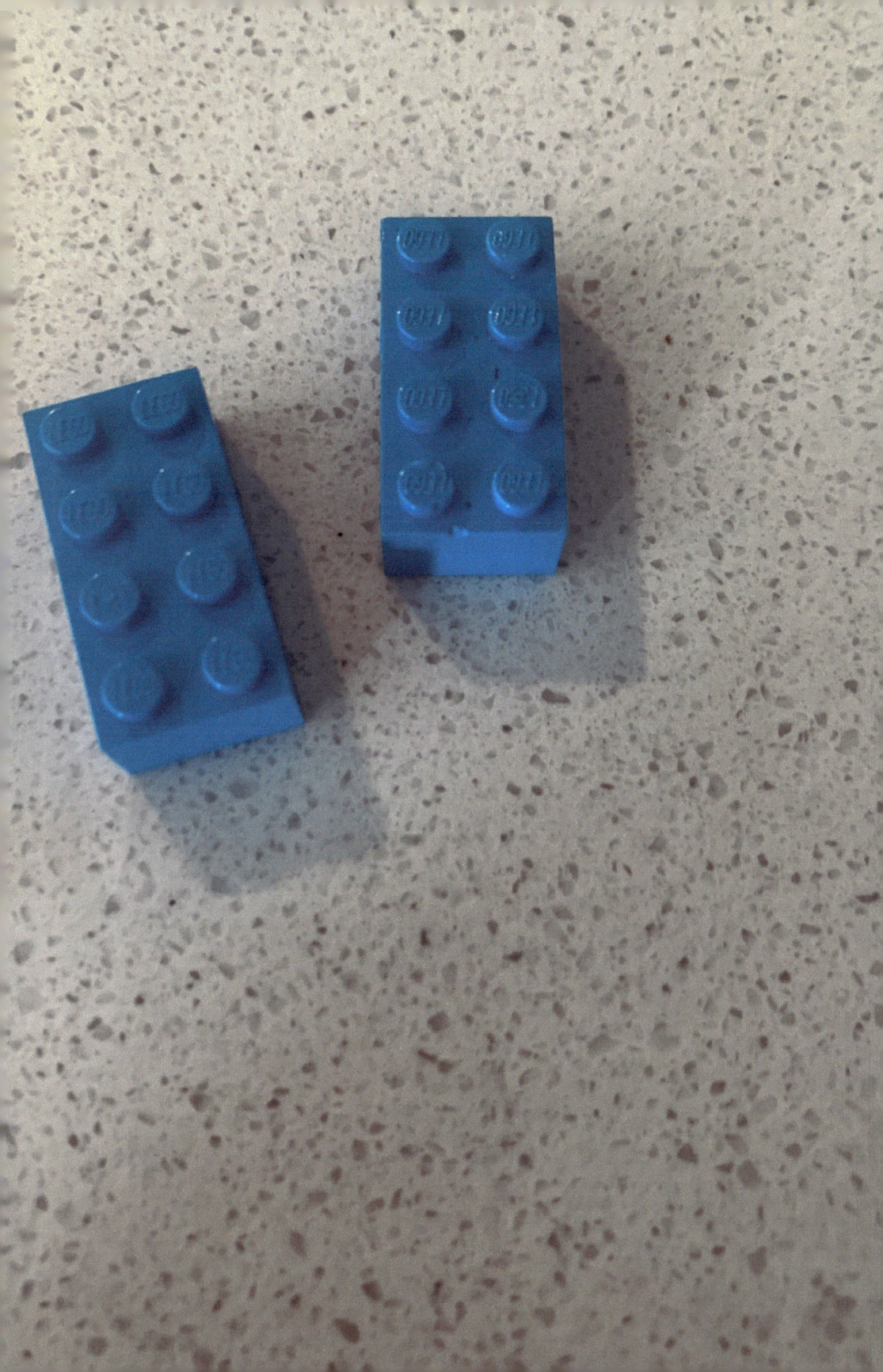

THREE

3

FOUR

FIVE

SIX

SEVEN

EIGHT

NINE

9

TEN

10

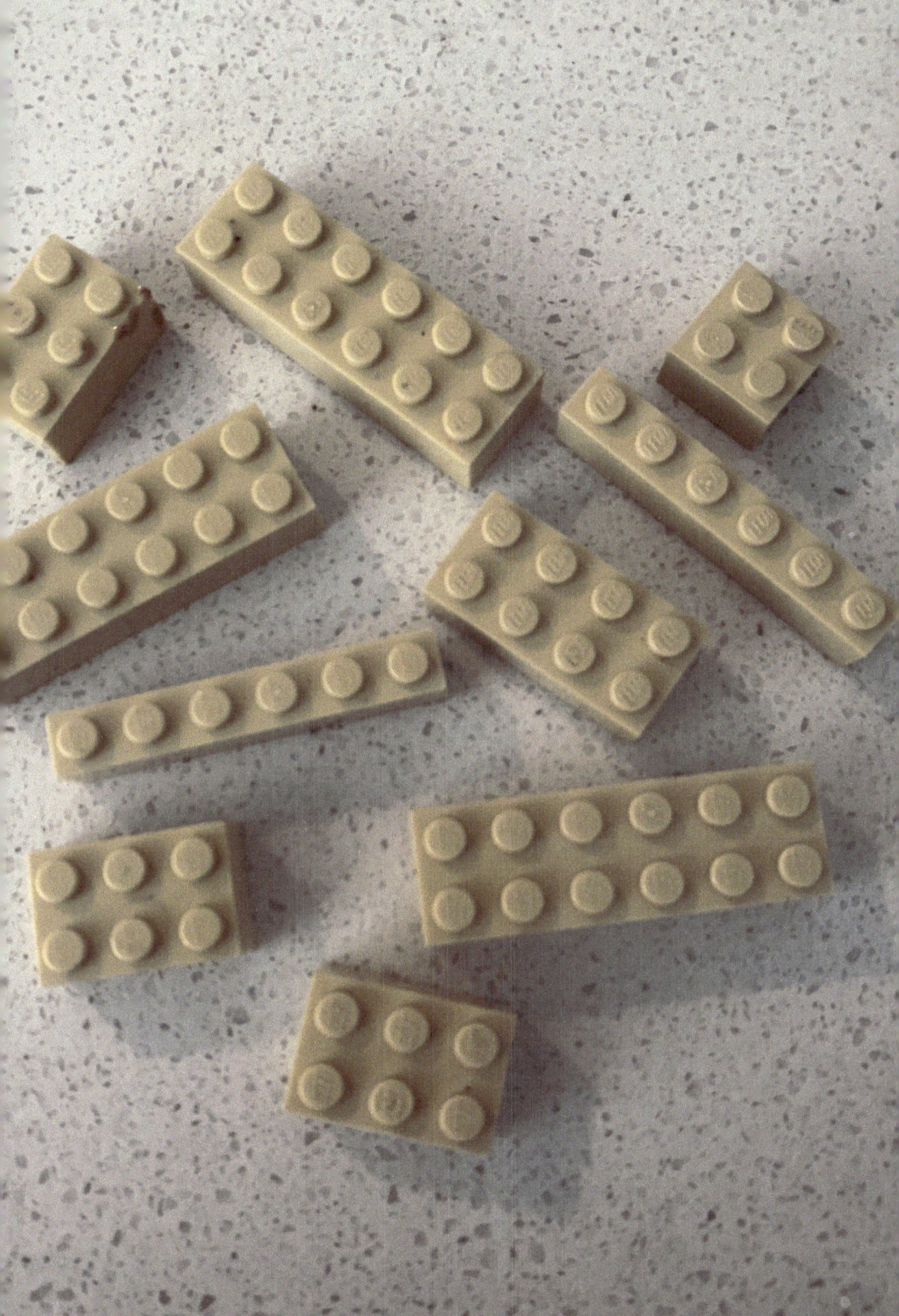

About The Author

I am an inspired writer. I enjoy what I am writing in these little kids books that I am recently writing . I like to see other kids enjoy what they are reading.

Books I wrote and has been published:
Our move to Florida
Unicorn Tea Party
Alice and her pink mirror!
Hear, Touch, See, Smell and Taste!
I Am!

www.ingramcontent.com/pod-product-compliance
Lightning Source LLC
Chambersburg PA
CBHW040351220526
45473CB00009B/2846